*For my son Balder
with love*

The cure for juvenile diabetes

*One is cured of juvenile diabetes if daily insulin injections are
no longer necessary and no other medication is needed*

Leo Rogier Verberne

The cure for juvenile diabetes

(type 1 diabetes)

2016

The cure for juvenile diabetes

Colophon

© 2016   Leo Rogier Verberne
ISBN/EAN   978-90-818362-9-6
Printing   Lulu Press Inc. USA
Internet   www.juvenile-diabetes-cure.org

Contents

Personal interest

One of our children developed juvenile diabetes. He was 15 years old at the time. It was a smack in his face and ours and it urged him to daily insulin injections for the rest of his life.
This confrontation raised several questions:
What is the cause of juvenile diabetes and how does it develop?
How can long-term complications be avoided?
Will it ever be curable?

Following my retirement, I had ample time to focus on this matter. As an internist for horses, I am familiar with literature on internal diseases. And that is what diabetes is. But more than anything else, I am the father of a son with juvenile diabetes.

Sint-Michielsgestel, June 2016
Leo Rogier Verberne

1. The way juvenile diabetes develops

Juvenile diabetes (type 1 diabetes or T1D) is the more severe form of diabetes that is usually diagnosed in children and teenagers, though it can become overt at any age (2).

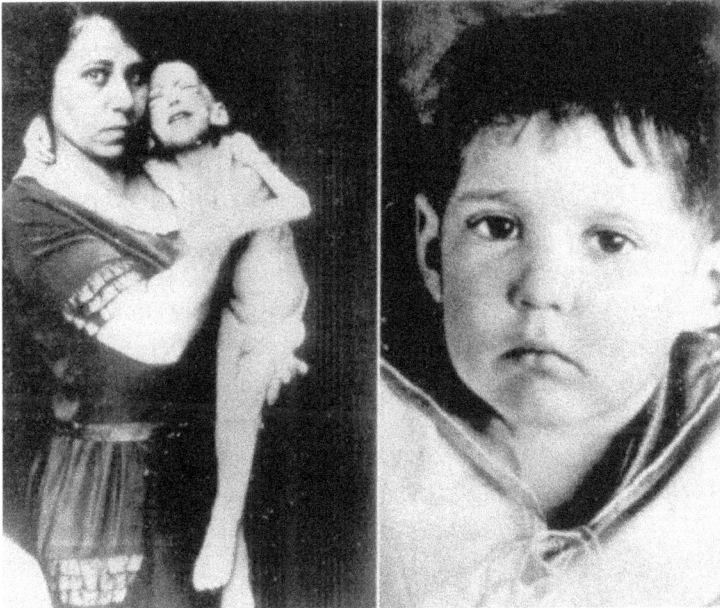

In 1922, Leonard Thompson was the first person to be treated with insulin. He was 14 years old at the time. On the right: the same boy six months later (6).

Case
Ms J., 57 years old, suffers from thirst and fatigue to an increasing extent. She is no longer able to carry out her work activities as an accountant. The numbers start to dance before her eyes after only a few hours; she cannot concentrate and as the afternoon progresses, she feels dead tired. These symptoms developed gradually over the course of a few weeks. She has difficulty sleeping because of her

thirst and because she has to urinate often. During the past week, she has had to drink at least a liter every night. Upon inquiry, it was found that she has always been in good health. According to her, there is no history of diabetes in her family.
She does not appear to be ill, but she does seem severely tired.
Her body weight is 63 kg at a height of 1.69 m. The fasting plasma glucose level of the blood is 22 mmol/l (normally < 6,1). In view of her age, she is diagnosed with type 2 diabetes and treated with 1 mg of glimepiride per day. But the symptoms worsen and despite an increased drug dosage, the patient becomes extremely nauseous and can no longer keep any food or fluids down.
Being taken to the emergency ward of a hospital, she is found to be in keto-acidosis. The presence of antibodies in her blood that target GAD antigens, reveals the diagnosis juvenile diabetes. Thereupon the medical treatment is changed from glimepiride tablets to daily insulin injections. If type 1 diabetes becomes overt at an age over 40 years, it is often termed LADA (Latent Auto-immune Diabetes of Adults). But that term is controversial and it could be better referred to as type 1 diabetes, coming about at a later age (5).

Heredity
In T1D patients, several gene mutants in the DNA, up to a maximum of 19, code for deviations in the immune system (3). As a result, derailed T lymphocytes mistakenly detect the insulin producing β cells in the pancreas as cancer cells or infected body cells and subsequently shut them down. Which is called an auto-immune reaction.
That terminates monitoring of the blood glucose concentration and insulin production. As a result, the blood glucose level rises above normal limits. So the ultimate cause of juvenile diabetes is in the DNA, which makes it a hereditary disease. It explains how inbreeding increases the incidence of T1D in closed populations, as in remote regions or on islands.

The way juvenile diabetes develops

Heel prick
Detection of deviating gene mutants in the DNA can be done as early as the heel prick in babies. Which enables intervention in the early development of juvenile diabetes, long before the disease becomes overt. Such an intervention is now within reach (see next chapter). But juvenile diabetics, after being cured in that way, still have the same gene mutants, that keep coding for deviations in the immune system. So the production of derailed T lymphocytes continues and the need for repeated interventions at a later age is to be expected.

Increasing numbers
Since the isolation of insulin in 1922, the number of juvenile diabetics is increasing worldwide. Before that moment, most of T1D patients died at a young age. But the availability of external insulin changed the former lethal disease into a discomfort of daily injections. As a result, juvenile diabetics could lead relatively normal lives, reached adulthood and created offspring. So their number has been increasing ever since. Moreover, the rise of the percentage of T1D patients in a population means an increasing chance for a juvenile diabetic to meet another T1D patient as sexual partner. That enlarges the risk of juvenile diabetes in their children about fourfold: from about 5% with one diabetic parent to ± 20% (3).

Auto-immune reaction
The regular immune system defends the body against diseases, detecting bacteria, viruses, parasitic worms, toxins and cancer cells as pathogens for elimination (8). But in juvenile diabetics, a derailed part of the immune system considers normal β cells in the pancreatic islands as pathogens and subsequently destroys them. That auto-immune reaction is crucial, both to the development of juvenile diabetes as well as to its cure.

The cure for juvenile diabetes

Because several gene mutants in the DNA code for deviations in the immune system, there are various ways to develop T1D and at different speeds. But sooner or later pancreatic β cells are shut down, leading to hyperglycemia and juvenile diabetes.

Autopsy

In post mortems of 45 juvenile diabetic organ donors with various T1D durations, up to 56 years, the pancreas was examined (1). Inflamed pancreatic islets appeared typically dominated by autoreactive CD8 T cells, causing CD8 insulitis in all T1D patients. Other leukocytes were more or less incidentally included. In 27 controls, derived from non-diabetics, T2D patients and gestational diabetics, no CD8 insulitis was found. So islet autoreactive (derailed) CD8 T cells are the key-evildoers in the shutdown of pancreatic β cells. Other leukocytes predominantly help to remove the killed β cells. In an overview of the literature about islet autoreactive CD8 T cells in type 1 diabetes, Roep suggests that such CD8 T cells might be suitable targets for immune intervention therapy in T1D (4).

Residual β cells

A second result of post mortems on juvenile diabetics: 'In a selective cohort of T1D patients, with disease durations of more than 50 years, many cases maintained a residual pool of functional β cells at the time they deceased' (1). So, even after several decades of disease, residual β cells can restart insulin production in many of the juvenile diabetics. Provided that all derailed CD8 T cells have been previously eliminated with efficient therapy. If the residual β cells sufficiently regenerate, then neither the supply of insulin nor a transplant of stem cells (as extra β cells) will be needed to keep the blood glucose level within normal limits. Thus an effective elimination of autoreactive CD8 T cells will cure juvenile diabetes, even without other medical interventions, in many T1D patients.

Islet autoreactive CD8 T cells

CD8 T cells (Cluster of Differentiation 8) are part of the immune
system. The prefix T refers to the thymus, a lymphoid organ that
matures certain lymphocytes (9). It is situated inside the chest between the heart and the sternum. CD8 T cells are also known as cytotoxic T cells, cytotoxic T lymphocytes or killer T cells (7). They reg-
ularly shut down cancer cells, infected body cells (particularly with
viruses) or cells that are damaged in other ways. However, if juvenile
diabetic gene mutants in the DNA code for a derailed maturing of
certain lymphocytes in the thymus, then islet autoreactive CD8 T
cells will develop, that shut down normal β cells in the pancreatic
islets, as if they were cancer cells or infected body cells.

Thymus

The thymus enlarges during childhood, reaching its maximum weight
(20 to 37 g in humans) by one's puberty (9). From then on it starts
shrinking. So an individual's supply of all kinds of matured T cells is
built up early in life. By the age of 75, the residual thymus weight is
only 6 grams, predominantly consisting of adipose tissue (9). Therefore, in elderly people, the loss of immune functions and increased
susceptibility to infection and cancer are linked to the reduced function of the thymus. However, the early shrinking of that organ is an
advantage for juvenile diabetics. After being successfully treated for
T1D, the gene mutants in the DNA keep on coding for deviations in
the immune system. But with increasing age, the thymus will hardly
be able to respond to that coding and will stop producing islet autoreactive CD8 T cells. Thus the need for repeated elimination of those
T cells in T1D patients will diminish after puberty.

Conclusions

1. Juvenile diabetes is a hereditary disease: DNA analysis in babies
   enables early detection of the deviating gene mutants for T1D.

2. Several gene mutants, up to a maximum of 19, code for deviations in the maturing processes of lymphocytes in the thymus, leading to the production of islet autoreactive CD8 T lymphocytes.
3. Islet autoreactive CD8 T cells mistakenly detect β cells in the pancreatic islets as cancer cells and shut them down, which is called an auto-immune reaction, leading to juvenile diabetes.
4. Elimination of islet autoreactive CD8 T cells will stop the auto-immune reaction in the pancreatic islets, thus stopping the shutdown of β cells.
5. Many juvenile diabetics have a residual pool of functional β cells, even after a disease duration of more than 50 years; thus effective elimination of islet autoreactive CD8 T cells will cure juvenile diabetes in many cases, even without other medical interventions.

References

1. Coppieters KT, Dotta F, Amirian N, Campbell PD, Kay TWH, Atkinson MA, Roep BO, von Herrath MG (2012).
Demonstration of islet-autoreactive CD8 T cells in insulitic lesions from recent onset and long-term type 1 diabetes patients.
J Exp Med. 2012 Jan16; 209 (1): 51-60.
doi: 10.1084/jem.20111187. Epub 2012 Jan 2
2. Diabetes Research Institute Foundation (2016).
What is Type 1 Diabetes?
http://www.diabetes research.org/what-is-type-one-diabetes
3. Hes FL en Breuning MH. *Klinische genetica.*
In: Interne geneeskunde. eds. Stehouwer, Koopmans en van der Meer. 14^e druk (2010); ISBN 978-90-313-7360-4; p 75-97
4. Roep BO (2008). *Islet autoreactive CD8 T-cells in type 1 diabetes.*
Diabetes 2008 May; 57(5): 1156-1159
http://diabetes.diabetesjournals.org/content/57/5/1156

5. Tack CJ en Stehouwer CDA. *Diabetes mellitus*
In: Interne geneeskunde. eds. Stehouwer, Koopmans en van der
Meer. 14^e druk (2010); ISBN 978-90-313-7360-4; p 842-843
6. Wientjens WHJM (2008). *Diabetes ... Nou en?*
zeventig jaar belevenissen.
Novo Nordisk BV; ISBN 978-90-804452-7-7; p 15
7. Wikipedia.en (2016). *Cytotoxic T cell*
https://en.wikipedia.org/wiki/Cytotoxic_T_cell
8. Wikipedia.en (2016). *Immune system*
https://en.wikipedia.org/wiki/Immune_system
9. Wikipedia.en (2016). *Thymus*
https://en.wikipedia.org/wiki/Thymus

2. The way to cure juvenile diabetes

Big business
In 2016, the estimated number of people suffering from diabetes mellitus worldwide is 422 million (3). The proportion of juvenile diabetics is estimated at 5% (2) to 10% (3). Thus between 20 and 40 million humans suffer from T1D, their lives depending on daily insulin injections. And their number is growing with increasing speed. The only option is to administer insulin, there is no cure for juvenile diabetes. The research into such a cure is big business. For example, in 2007 the pharmaceutical company Eli Lilly licensed teplizumab as a promising drug for curing juvenile diabetes, with a $ 41 million upfront and a billion dollars pledged for milestones (15). A real T1D cure will phase out daily insulin injections. Which will be a landslide, both in the pharmaceutical field as well as in health care.

Gene therapy
The cause of juvenile diabetes is in the DNA. Its hereditary predisposition varies up to a maximum of 19 deviating gene mutants. Thus to tackle this disorder at its source, gene therapy seems indicated (9). However, recent corrections made in the genetic makeup of embryo's by Chinese researchers, did not improve 'bad genes', while disturbing a lot of 'good genes'(1). So gene therapy, as a cure for T1D, seems not to be for the near future.

Stopping the auto-immune reaction
Another intervention to cure juvenile diabetes might be in stopping the auto-immune reaction in the pancreatic islets. Nowadays most auto-immune diseases are treated by a nearly general suppression of the immune system, for example by prednisone or ciclosporine. However, that effect is only temporary so that lifelong medication is necessary. But those drugs also have serious toxic side effects, while general immune depression involves a risk of various infections.

Therefore, the *selective* shutdown of derailed immune cells is preferable. To that end, monoclonal antibodies are in use. They shut down certain immune cells by targeting specific receptors on their cell membrane. However, they do not destroy these cells. So the beneficial effect of monoclonal antibodies is also temporary, and again lifelong medication is needed. Most of them also have severe side effects. Thus the advantage compared to general immune suppression is only in the smaller risk of secondary infections.

CD8 monoclonal antibody

In juvenile diabetes, the selective shutdown of islet autoreactive CD8 T cells requires a monoclonal antibody that targets CD8 receptors on the membrane of those T cells. Monoclonal antibodies are produced by cloning specific parent cells in transgenic mice or by phage display (11). As these are living organisms, the resulting monoclonal antibody is called a 'biological'. But available CD8 biologicals are only licensed for diagnostic laboratory tests (4). They have not been approved for therapeutic use in patients (10). To that end, a fully human biological is preferable, because it does not invoke an immune response against foreign proteins in humans (11). 'Fully human' means that the biological has been derived from a human parent cell. In the case of juvenile diabetes, that parent cell must be an autoreactive CD8 T lymphocyte. Which can be found in the inflamed pancreatic islands of T1D patients in post mortems. The monoclonal antibody against such CD8 T cells could be named *oktolimumab*. The prefix *okto* (eight) referring to the CD8 receptor to which the antibody affixes. According to the nomenclature of monoclonal antibodies, *lim* indicates the targeted lymphocytes; the letter *u* is for the fully human nature of the antibody and *mab* stands for monoclonal antibody (13).

Some companies with experience in the production of fully human monoclonal antibodies:
- *AbbVie Inc., Illinois, USA:*
  adalimumab (Humira®)
- *Agensys Inc.(Astellas Pharma Inc.), Los Angeles, USA:*
  enfortumab
- *Torrent Pharmaceuticals Ltd, Ahmedabad, Gujarat, India:*
  adalimumab
- *Zydus Cadila, Ahmedabad, Gujarat, India:*
  adalimumab (Exemptia®)

ADC

A recent technique to reduce the side effects of toxic drugs is by engineering an ADC (Antibody-Drug Conjugate) i.e. a monoclonal antibody conjugated to a chemotherapeutic (cytotoxic) drug (7). Nowadays, ADC's are engineered for cancer therapy. The monoclonal antibody specifically targets certain cancer cells, which subsequently are destroyed by the chemotherapeutic drug. By combining the unique targeting capabilities of a biological with the cancer killing ability of a cytotoxic drug, an ADC allows sensitive discrimination between healthy and diseased tissue (7). Thanks to the selective release of the cytotoxic drug, its dosage can be considerably lower and side effects are substantially diminished. Thus it is possible to use drugs 100 to 1000 times more cytotoxic for cancer cells than drugs in use today (6).

Vedotin

A potent cytotoxic drug is vedotin (12). It is derived from a marine mollusc. After invading a cell, it splits off monomethyl auristatin E (MMAE), which inhibits cell division and causes cell death. Because of its toxicity, MMAE cannot be used as a drug by itself. But when conjugated (as vedotin) to a monoclonal antibody, it selectively kills the target cells, meanwhile avoiding most of its toxic side effects. Therefore, the link between the biological and vedotin must be stable, so that it is not cleaved before it has entered the targeted cancer cells.

ADC and cancer

Anno 2016, only two ADC's have been approved for cancer therapy so far: *Brentuximab-vedotin* (Adcetris®) and *Trastuzumab-emtansine* (Kadcyla®). Adcetris® is in clinical use for classical Hodgkin lymphoma (HL) and systemic anaplastic large cell lymphoma (8). The monoclonal antibody brentuximab targets the cell-membrane receptor CD30 of these rapidly dividing cells. Inside these cells, it splits off MMAE which kills the cancer cells. Kadcyla® is approved for HER2-positive metastatic breast cancer (16). The biological trastuzumab targets the receptor HER2, which is over-expressed in that type of breast cancer cells. The cytotoxic agent emtansine subsequently destroys these cells.

Some companies with experience in the production of ADC's:
- Lonza Group Ltd (F. Hoffmann-La Roche AG), Basel, Switzerland:
  trastuzumab-emtansine (Kadcyla®)
- Pfizer Inc BioTherapeutics R&D, Cambridge, Massachusetts, USA:
  human F8 antibody-interleukine-10 (Dekavil)
- Millennium Pharmaceuticals Inc.(Takeda Oncology Company),
  Cambridge, Massachusetts, USA and
- Seattle Genetics, Bothell, Washington/Seattle, USA:
  jointly developing brentuximab-vedotin (Adcetris®)
- Synaffix, Oss, The Netherlands

ADC and juvenile diabetes

In juvenile diabetes, islet autoreactive CD8 T cells can be considered cancer cells. So an ADC that selectively kills these cells requires a monoclonal antibody that specifically targets them. Such a biological is (or will be soon) oktolimumab. The conjugated cytotoxic drug could be vedotin. This newly to be engineered ADC *oktolimumab-vedotin* must target and destroy islet autoreactive CD8 T cells, leaving the rest of the immune system intact. As a consequence, the auto-immune reaction in the pancreatic islets will stop, enabling recovery of residual β cells and the restart of insulin production. In patients with insufficient revival of β cells, implantation of stem cells (as extra β cells) might complete the cure for juvenile diabetes.

Control of the cure
If the ADC *oktolimumab-vedotin* does indeed destroy all islet autoreactive CD8 T cells in the patient's body (possibly after repeated doses), then residual and still functional β cells will regain control of the blood glucose concentration and they will restart insulin production, thus gradually banning daily insulin injections.
Outcome measures for this ADC therapy will be:
- decrease of anti-GAD in the patient's blood, indicating the release of the auto-immune reaction;
- increase of the C-peptide level, reflecting the repair of β-cell function.
Together, both laboratory tests control the cure for juvenile diabetes and monitor a possible relapse.

Relapse
Even a complete cure for juvenile diabetes will not necessarily be permanent, because the aberrant gene mutants in the DNA continue to code for faulty maturing of CD8 T cells in the thymus. However, from puberty onwards, the thymus is shrinking. Thus the older the T1D patient is at the time of a successful immune intervention therapy with *oktolimumab-vedotin*, the less active the thymus will then be and the smaller the chance of a relapse of juvenile diabetes.

The way to cure juvenile diabetes

Diabetic dogs

The approval of *oktolimumab-vedotin* as a new therapeutic drug in juvenile diabetes, is a big and promising challenge. A license by the FDA (Food and Drug Administration) requires a lot of research, involving laboratory animals as well as T1D human patients. Diabetic mice and rats are extensively used in the research into juvenile diabetes. But the predictive significance of those experiments for T1D in humans was found to be disappointing.

To that end, *diabetic dogs* might be helpful. So far they are beyond the scope of researchers in human diabetes. Diabetes mellitus is one of the most common endocrine disorders in dogs (5), having a prevalence of 0.3-0.6 %. In many dogs the disease is similar to human type 1 diabetes, which is caused by auto-immune destruction of β cells in genetically predisposed individuals. Likewise, certain breeds of dogs are predisposed to diabetes. Since most of these animals are middle-aged to elderly at the time of diagnosis, canine type 1 diabetes seems to correspond best to the subgroup of human type I diabetes termed LADA (Latent Auto-immune Diabetes in Adults) (5).

Pilot study

As a pilot study, the ADC *oktolimumab-vedotin* could be tested for its curing capacity in diabetic dogs. Because they will be of various breeds, their size and weight can considerably differ. Therefore the ADC should be dosed in mg/kg body weight. Subsequently, doses could be increased step-by-step in, for example, monthly intervals. That might yield preliminary insight into the curing capacity and the side effects of the ADC. If one or more dogs could be cured that way, it would give an enormous boost to the research into the cure for human juvenile diabetes. Likewise, experiments in dogs by Banting and Best preceded the treatment of human diabetic patients with *insulin* in 1922. Banting and Macleod were jointly awarded the 1923 Nobel Prize in Physiology or Medicine (14).

The cure for juvenile diabetes

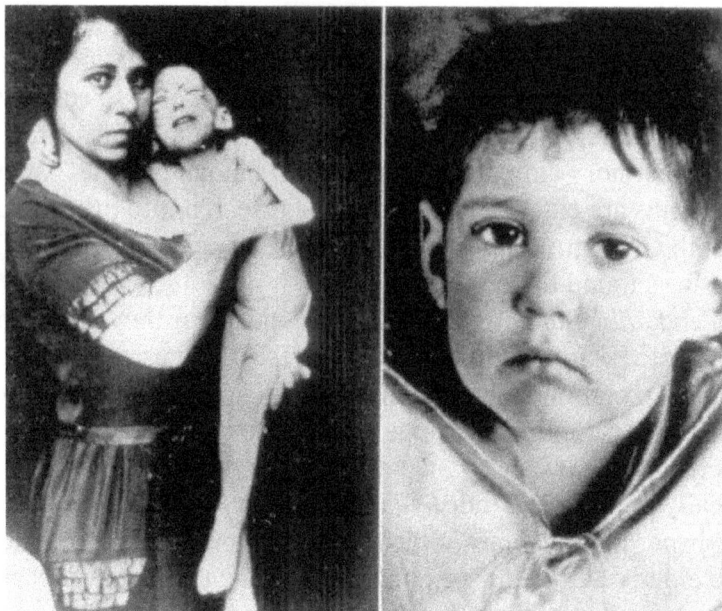

*After isolation of insulin to treat juvenile diabetes,
a second giant step might be to cure type 1 diabetes
by oktolimumab-vedotin*

Conclusions
1. Anno 2016, there are between 20 and 40 million juvenile diabetics worldwide; so research into the cure for T1D is big business.
2. Gene therapy is the preferable remedy when tackling a hereditary disease like juvenile diabetes at its source; but that therapy is still in its infancy.
3. General suppression of the immune system stops auto-immune reactions, but it evokes serious infection risks and side effects; and its effect is temporary so that a lifelong medication is needed.

4. Selective suppression of certain immune cells by monoclonal anti-bodies also has a temporary effect; but it has the advantage, compared to general immune suppression, of a smaller infection risk.
5. A fully human monoclonal antibody against islet autoreactive CD8 T cells (oktolimumab) will selectively target those derailed T cells, which shut down β cells in the pancreatic islets.
6. The ADC *oktolimumab-vedotin* will selectively destroy islet auto-reactive CD8 T cells, thus stopping the shutdown of β cells; as a result, residual β cells will restart insulin production.
7. Once there is a cure for juvenile diabetes, the risk of a relapse diminishes with age: the older the T1D patient at the time he/she is cured, the smaller the risk of a relapse; because the thymus starts shrinking from puberty onwards.
8. Diabetic dogs might serve to test the curing capacity of drugs for juvenile diabetes; one or more cured dogs would give an enormous boost to the research into the cure for human T1D.

References
1. Borst P (2015). *Eugenetica revival.*
   Column NRC Handelsblad 16 mei 2015
2. Diabetes Research Institute Foundation (2016).
   What is Type 1 Diabetes?
   http://www.diabetesresearch.org/what-is-type-one-diabetes
3. Koning E de (2016). *Hubrecht college.*
   Amsterdam, 6th of April 2016
4. LSBio (2016). *Anti-CD8 Antibodies.*
   https://www.lsbio.com/search?q=CD8+anti-bodies&sid=iuj5wd
   7262
5. Reusch CE, Robben JH, Kooistra HS (2010). *Diabetes mellitus in Dogs.* In: Clinical Endocrinology of Dogs and Cats. eds. Rijnberk and Kooistra. 2nd ed. (2010); ISBN 978-3-89993-058-0; p 161-167

6. Seattle Genetics (2016).
   Advancing industry-leading ADC technology.
   http://www.seattle-genetics.com/adc_technology
7. Wikipedia.en (2016). *Antibody-drug conjugate*
   https://en.wikipedia.org/wiki/Antibody-drug_conjugate
8. Wikipedia.en (2016). *Brentuximab vedotin*
   https://en.wikipedia.org/wiki/Brentuximab_vedotin
9. Wikipedia.en (2016). *Gene therapy*
   https://en.wikipedia.org/wiki/Gene_therapy
10. Wikipedia.en (2016). *List of therapeutic monoclonal antibodies*
    https://en.wikipedia.org/wiki/List_of_therapeutic_monoclonal_
    antibodies
11. Wikipedia.en (2016). *Monoclonal antibody.*
    https://en.wikipedia.org/wiki/Monoclonal_antibody
12. Wikipedia.en (2016). *Monomethyl auristatin E*
    https://en.wikipedia.org/wiki/Monomethyl_auristatin_E
13. Wikipedia.en (2016). *Nomenclature of monoclonal antibodies.*
    https://en.wikipedia.org/wiki/Monoclonal_antibody
14. Wikipedia.en (2016). *Insulin.*
    https://en.wikipedia.org/wiki/Insulin
15. Wikipedia.en (2016). *Teplizumab*
    https://en.wikipedia.org/wiki/Teplizumab
16. Wikipedia.en (2016). *Trastuzumab emtansine*
    https://en.wikipedia.org/wiki/Trastuzumab_emtansine

Author

Leo Rogier Verberne

1943 born in Helden-Panningen (The Netherlands)
1962 gymnasium in Eindhoven
1970 veterinary surgeon examination; clinic for internal diseases,
faculty of veterinary medicine in Utrecht
1973 clinic for veterinary surgery,
faculty of veterinary medicine in Utrecht
1978 laboratory for physiology, medical faculty V.U. Amsterdam
1983 registration as Medical Physiologist
1986 obtained degree of PhD medicine

1984 medical practice for agricultural animals and pets in Hintham
(near Den Bosch)
1987 medical practice for agricultural animals and horses in
Berlicum
2000 medical practice for agricultural animals and horses in Oss
1994 registration as specialist Equine Internal Medicine
2000 certification as cattle veterinary specialist
2001 certification as horse veterinary specialist
2003 retired veterinarian and retired horse veterinary specialist

Samenvatting

1. Hoe ontstaat jeugddiabetes?

Jeugddiabetes is een erfelijke aandoening. De aanleg kan al worden vastgesteld bij baby's door DNA- onderzoek van bloed dat wordt verkregen bij de hielprik. Verschillende gen-mutanten in het DNA, tot een maximum van 19, coderen voor het ontstaan van afwijkingen in het immuunsysteem die tot jeugddiabetes leiden. Dat verloopt via de thymus, een orgaan in de borstholte, waar afwijkingen ontstaan in het rijpingsproces van de lymfocyten. Er worden dan autoreactieve CD8 T-lymfocyten gevormd die in de alvleesklier een auto-immuun-reactie veroorzaken die β-cellen vernietigt. Uitschakeling van deze autoreactieve CD8 T-cellen zal de auto-immuunreactie in de alvlees-klier stoppen en jeugddiabetes genezen. Uit secties blijkt dat bij veel jeugddiabeten, zelfs na meer dan 50 jaar type 1 diabetes, nog een reservevoorraad van functionele β-cellen in de alvleesklier aanwezig is. Regeneratie van deze cellen zal de productie van eigen insuline weer op gang brengen.

2. Hoe jeugddiabetes kan worden genezen

Anno 2016 zijn er naar schatting 20 tot 40 miljoen jeugddiabeten wereldwijd. Dus het onderzoek naar de genezing is 'big business'. Om een erfelijke aandoening te genezen, lijkt gentherapie de aange-wezen behandeling. Maar deze tak van de medische wetenschap staat nog in de kinderschoenen. De behandeling van auto-immuunziekten bestaat thans uit een *algemene suppressie* van het immuunsysteem d.m.v. prednison of ciclosporine. Dat stopt auto-immuunreacties, maar het effect is slechts tijdelijk en daardoor is levenslange medica-tie noodzakelijk. Dat veroorzaakt een groot risico op allerlei infecties en er zijn ernstige, toxische bijverschijnselen.
Selectieve suppressie van alleen bepaalde immuun-cellen is mogelijk d.m.v. monoclonale afweerstoffen. Ook hun effect is slechts tijdelijk en ook die hebben ernstige bijwerkingen. Maar het voordeel boven algemene immuun-suppressie is een kleiner risico op infecties.

The cure for juvenile diabetes

Een monoclonale afweerstof tegen autoreactieve CD8 T-cellen voor therapeutisch gebruik bij jeugddiabeten is er nog niet. Zo'n afweerstof moet selectief hechten aan deze T-cellen. Als ouder-cel bij het klonen zou daarom een autoreactieve CD8 T-cel moeten dienen zoals die bij overleden jeugddiabeten in de alvleesklier worden aangetroffen. Conform de regels voor naamgeving van monoklonale afweerstoffen, zou deze 'oktolimumab' genoemd kunnen worden. Door dit oktolimumab te koppelen aan een krachtige cytotoxische stof, bijv. vedotin, ontstaat het ADC (Antibody-Drug Conjugate) *oktolimumab-vedotin* dat de auto-reactieve CD8 T-cellen in de alvleesklier van jeugddiabeten selectief kan vernietigen. Dit betekent de genezing van de jeugddiabeet, doordat de overgebleven β-cellen de productie van insuline dan zullen hervatten. Alleen als de regeneratie van deze β-cellen tekortschiet, zal een aanvullende behandeling met stamcellen (als extra β-cellen) nodig zijn. Na een succesvolle ADC-therapie bestaat op termijn het risico van terugval, omdat de afwijkende genmutanten in het DNA aanwezig blijven. Dat risico wordt echter steeds kleiner naarmate de patiënt ouder is, doordat de thymus al vanaf de puberteit kleiner en minder functioneel wordt. Dus zullen op latere leeftijd geen nieuwe auto-reactieve CD8 T-cellen meer aangemaakt worden.

Bij veel honden met spontane diabetes mellitus lijkt de aandoening op type 1 diabetes van de mens. Daarom zouden deze honden goede proefdieren zijn om een nieuw medicijn te testen voor de genezing van T1D, zoals het ADC *oktolimumab-vedotin*. Als één of enkele honden daarbij genezen, zal dat het onderzoek naar de genezing van jeugddiabetes bij mensen een geweldige boost geven.